Transcranial Magnetic Stimulation
A Primer

ISBN 9798882972386

Dedication

To all my fellow researchers and clinicians who are also interested in TMS

Introduction

Transcranial Magnetic Stimulation (TMS) has undergone a significant transformation, transitioning from an investigative tool designed to probe the functioning of the central nervous system in both its normal and pathological states, to a therapeutic instrument with the potential to modulate neural activity in the treatment of various neuropsychiatric conditions. The sophistication and potential of TMS necessitate a comprehensive manual that encompasses its historical development, the fundamental scientific principles underpinning its operation, the regulatory framework governing its use, and a detailed discussion of its established and investigational therapeutic applications.

It is with these considerations in mind that the current volume has been compiled. Our intention is to provide an authoritative resource that lays a solid foundation for professionals and researchers interested in harnessing TMS within the fields of neurology and psychiatry. This book aims to equip its readers with the requisite knowledge to further investigate and apply TMS as a pioneering therapeutic modality, thereby contributing to the advancement of neurological and psychiatric care.

Contents

Foreword

In 1990, I arrived in Bethesda, Maryland, United States of America, to begin a 3-year period as a Visiting Fellow at the National Institutes of Neurological Disorders and Stroke, under the supervision of Dr. Mark Hallett. It was then that I first encountered the technique of transcranial magnetic stimulation (TMS). At that time, this technique was only available in England and the United States, exclusively for research purposes.

As we will see later in this text, the motivation for developing a non-invasive and painless procedure for stimulating the central nervous system was its potential use as a diagnostic tool in neurological diseases, such as multiple sclerosis. In this demyelinating disease, tests such as visual, auditory, and somatosensory evoked potentials were already well established, capable of detecting demyelination in the sensory pathways of the central nervous system; however, the study of motor pathways would require a technique for stimulating the motor cortex, that is, a method for obtaining motor evoked potentials.

The ease of obtaining motor responses to cortical stimulation painlessly in awake volunteers naturally led to the use of TMS in studies of the normal physiology of motor pathways. Furthermore, by the late 1980s and early 1990s, many studies were being conducted in animals, demonstrating an unsuspected degree of central nervous system plasticity in adults in response to various manipulations such as limb amputation, artificial syndactyly, motor learning, etc. These studies focused on the central sensory systems due to the technical ease of their study in animals... and with TMS, it was now extremely easy to map, for example, the motor somatotopic representation in humans in a non-invasive way! Therefore, many findings of acute or chronic brain plasticity demonstrated in sensory systems in animals could be complemented with corresponding studies in the human motor system.

During the performance of numerous experimental TMS protocols in humans, it became clear that the application of magnetic pulses repetitively could modulate the excitability of the motor cortex for time intervals that persisted after the cessation of the stimuli. This phenomenon was similar to that of LTP (long-term potentiation) and LTD (long-term depression) already known from *in vitro* studies on synaptic modulation with repetitive electrical stimuli.

The possibility of modulating cortical excitability with TMS, and the dissemination of studies suggesting an imbalance in the excitability of the prefrontal cortices in patients with major depression, led to the experimental hypothesis that TMS, applied repetitively and in sessions, could modulate prefrontal cortical excitability, restoring the balance between cerebral hemispheres and improving the depressive symptoms. As we will see in this text, several studies were conducted, including multicenter controlled studies, confirming this hypothesis and culminating in the approval of this new therapeutic modality by the FDA in the United States and by the Federal Council of Medicine in Brazil.

Repetitive transcranial magnetic stimulation (rTMS) is currently used primarily as a non-pharmacological treatment for depression. It has the disadvantages of being labor-intensive and requiring multiple patient visits to the clinic, but because it is not a systemic treatment, it can be used, for example, in pregnant women, for whom antidepressant drugs are contraindicated. Its main indication, however, lies in cases of depression refractory to drug therapy. Recently, a panel of European experts reviewed the clinical uses of rTMS, concluding that high-frequency stimulation of the left dorsolateral prefrontal cortex has level "A" evidence (proven efficacy) for the treatment of depression [51, 50].

Among the factors hindering the wider dissemination of this new therapeutic modality, we can highlight: the confusion with various "magnetic" products long associated with unscientific panaceas; the belief that it would be a new form of electroconvulsive therapy (popularly known as "shock therapy"); and simply the unawareness of this treatment, as it is relatively new (the first magnetic stimulator was built in 1985). However, the major obstacle to a wider diffusion of this revolutionary type of treatment lies in the difficulty encountered, both by patients and by healthcare professionals, in finding correct and reliable information in Portuguese, gathered in one place. There is some high-quality material in Portuguese, but with an academic bias and often emphasizing the use of the technique in research rather than in clinical practice.

The purpose of this manual is to provide healthcare professionals with a collection of updated and evidence-based information, which is indispensable for those intending to make clinical use of rTMS. Here, references and links are provided to documents covering medical, technical, ethical, and legal aspects of the clinical use of rTMS in patients with depression.

rTMS has already been approved for clinical use in depression by the Federal Council of Medicine, which conditioned the therapeutic application of the technique to the existence, in the care environment, of some minimum patient safety requirements. There are also international consensuses regarding desirable treatment protocols and essential physical characteristics of rTMS services, as well as their workflows. All these recommendations will be addressed here, with indications of extensive material for consultation. Furthermore, this new edition brings important updates, especially regarding the latest level of evidence of rTMS in various neuropsychiatric pathologies and new advances in the use of another modality of non-invasive neuromodulation, transcranial direct current stimulation (tDCS).

Preface

> I don't know what your destiny will be, but one thing I know: the only ones among you who will be really happy are those who will have sought and found how to serve. — Dr. Albert Schweitzer

Since the initial application of transcranial magnetic stimulation (TMS) in clinical environments, neurologists and other medical personnel in Brazil, and likely in other nations, have faced difficulties in accessing basic and regulatory information. Much of the knowledge acquired originated from industry brochures, presenting an inherent risk of bias. As one of the pioneering researchers to employ TMS technology, I recognized a responsibility to assist my peers by providing accurate physiological, technical, and regulatory information to support their endeavors in establishing TMS clinics. Following an extensive and meticulous compilation of essential and practical information, I undertook the task of writing this book. It is my hope that this text will serve as a valuable resource for colleagues embarking on the exploration of the therapeutic capabilities of this innovative tool.

Joaquim P. Brasil-Neto

Brasília, Brazil, March 22, 2024

1 A Brief History of TMS

1.1 A Brief History of Brain Stimulatiom

1.1.1 The early days of transcranial electrical stimulation

The medicinal use of electricity has a long history. As early as the 1st century AD, Scribonius Largus (1-50 AD) introduced therapy with electric fish (torpedoes) for the relief of gout and headache. Galen (130-c.201) also described the paralyzing and analgesic properties of the torpedo fish. The Arabic medical texts of Avicenna also indicated therapy with electric fish for the cure of headaches, melancholy, and to interrupt convulsive seizures [98].

In the Middle Ages, the powers of the torpedo fish were considered magical and supernatural. Only after much speculation about how a fish could paralyze its victims so quickly did Henry Cavendish (1731-1810) observe that the "sting" of the torpedo was similar to the shock produced by a Leyden jar (a primitive type of electrical capacitor) and built an experimental model of an electric fish composed of a series of Leyden jars housed in a leather case the size and shape of the fish [98].

Excitable animal tissue had been electrically stimulated around 1780 by Luigi Galvani (1737-1798) at the University of Bologna: using a Leyden jar, he demonstrated that its discharge was capable of inducing movement of a frog's leg, stimulating both the nerve and the muscle.

Galvani's nephew, Giovanni Aldini, conducted various experiments with galvanic stimulation of animals and human cadavers, obtaining impressive muscle contractions. The cadavers used were those of criminals condemned to death by the guillotine, as the same sentence ordered that the bodies be dissected afterward, supposedly to prevent their resurrection on Judgment Day.

In his experiments with freshly decapitated heads, Aldini found that the passage of an electric current, either through the ear and mouth or through the exposed

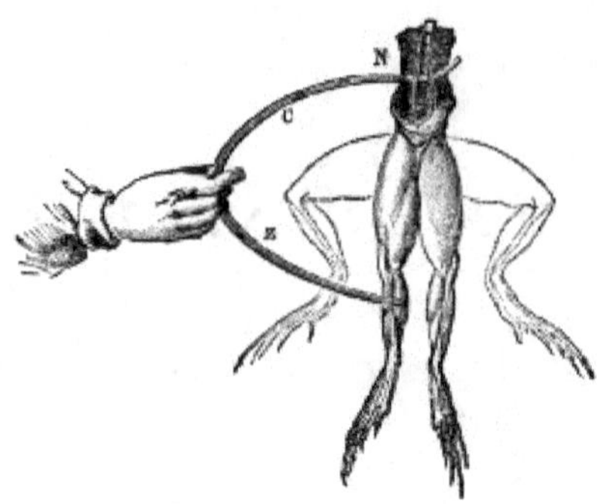

Figure 1.1: Illustration of one of Galvani's experiments

brain and mouth, produced various facial contractions. The fresher the head, the more easily these contractions occurred.

After his experiments with cadavers, Aldini began experimenting on volunteers and patients, applying the current from voltaic cells. After experimenting on himself, with electrodes in both ears or in one ear and the mouth, or on the forehead and nose, he experienced a strong reaction (*une fort action*), followed by prolonged insomnia that lasted for several days. He considered the experience very unpleasant but suggested that the changes produced in the brain could be beneficial in psychoses (*la folie*). Passing currents between the ears produced pain and violent convulsions, but he reported good results in patients suffering from melancholy. Aldini did not have instruments to measure the intensity of the applied currents (he only recorded the number of copper and zinc disks in the voltaic cell). Even in the adaptation of his electroconvulsive therapy technique by Ugo Cerletti in the twentieth century [14], there is no description of the current used, but only of the voltage:

> I decided to start cautiously with a low-intensity current of 80 volts for 1.5 seconds... The electrodes were applied again and a discharge of 110 volts was applied for 1.5 seconds.

It is not surprising that Cerletti's pioneering procedure resulted in convulsion, apnea, and cyanosis...

The first report of direct cortical electrical stimulation in living humans was by Bartholow in 1874. The experiment, which still raises ethical questions today, was performed on Mary Rafferty, a patient with exposure of the cerebral cortex secondary to cancerous erosion of the skull [35].

Throughout the eighteenth and nineteenth centuries, many experiments followed, confirming the susceptibility of nervous tissue to electrical stimulation, and in 1831, Michael Faraday discovered electromagnetic induction.

1.1.2 Magnetophosphenes and peripheral nerve magnetic stimulation

Neural tissue electrical stimulation can also be achieved by electromagnetic induction, as demonstrated by d'Arsonval, a physicist and physician, in 1896. He passed high-intensity electrical currents through coils around volunteers' heads and found that phosphenes, vertigo, and, in some cases, syncope were produced.

Today we know that phosphenes occur due to electromagnetic stimulation of the retina [29]. The phenomenon of magnetophosphenes became very popular when alternating current was being developed for commercial purposes and was also studied by other authors, such as Beer, Sylvanus Thompson, Dunlap, Walsh, Magnusson, and Stevens, in the first decade of the twentieth century. After three decades of being forgotten, magnetophosphenes were studied again by Barlow, who confirmed their retinal origin.

In 1965, Reginald Bickford , an electroencephalographer, described the stimulation of human peripheral nerves by a pulsed magnetic field [5]. He attempted to stimulate the brain, but although volunteers reported a sensation of movement in their legs, no objective movement was recorded. Bickford wrote, in 1965:

> Our interest in magnetic stimulation stems from the possibility that magnetic fields of sufficient intensity could stimulate cortical structures through the intact skull. Stimulation thus obtained could constitute a useful research tool with applications in the diagnosis of

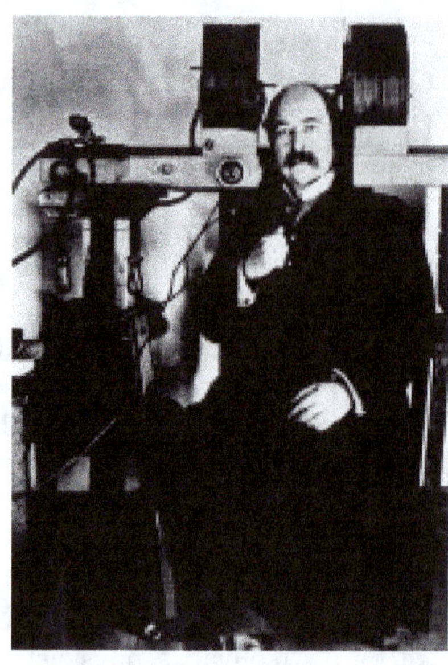

Figure 1.2: Sylvanus Thompson trying to magnetically stimulate his own brain

brain lesions and possibly in treatment.

We can say, therefore, that Bickford's words were prophetic, and in reality, the equipment he used suffered from the difficulty of bringing the coil close to the head, because the intensity of the magnetic field was good: its magnetic field threw a 2 kg metal cube to a height of 60 cm! Were it not for this technical problem, transcranial magnetic stimulation could have been invented 20 years earlier!

1.2 Transcranial magnetic stimulation

Merton, in 1981, after receiving a high-voltage electrical stimulator built by Morton, experimented on his own scalp to verify if he could electrically stimulate the brain. He was successful! Merton, Morton and Marsden also demonstrated that electrical stimuli could be applied to the spinal cord to produce muscle responses. However, the electrical stimuli were painful.

Figure 1.3: The first transcranial magnetic stimulator

1.2.1 The first transcranial magnetic stimulation of the human cortex

On February 12, 1985, at Queen Square Hospital in London, Barker magnetically stimulated Morton's brain with equipment built in Sheffield. The stimulation was completely painless. Thus, the first description of magnetic stimulation of the human brain was published in 1985 [3].

1.2.2 Clinical and research uses

In the following years, the possibility of non-invasive and painless stimulation of the human motor cortex was clinically explored for the study of central motor conduction time, using the technique of motor evoked potentials [57], and also as an experimental tool for studying the physiology and plasticity of the nervous system [97, 20, 18].

Already by 1994, however, it became clear that repeated magnetic stimuli applied to the human motor cortex were capable of producing changes in cortical excitability that persisted after the end of the stimulation session [70]. This finding was interesting as it was reminiscent of results obtained by electrical stimulation of the hippocampus and cerebellum in animals, which led to the description of long-term potentiation (LTP) and long-term depression (LTD) phenomena [7, 55]. One hypothesis for the mechanism of action of rTMS on the human motor cortex is the induction of phenomena similar to LTP and LTD.

However, actions on underlying cortical rhythms, alterations in neurotransmitter release, and actions on distant, subcortical circuits have also been postulated [51].

1.2.3 A pioneer study in depression

In 1996, based on accumulated evidence of dysfunction in prefrontal cortical areas in major depression [30], Pascual-Leone published a landmark article showing that repetitive transcranial magnetic stimulation (rTMS) applied over the left dorsolateral prefrontal cortex was able to improve severe cases of depression resistant to pharmacological therapy [70].

Since then, and based on the study of changes in cortical excitability that can be induced by rTMS [71], this form of stimulation, applied in repeated sessions, has been used in various neuropsychiatric disorders, always with the aim of durably exciting or inhibiting certain brain regions that are presumed to be involved in the pathophysiology of these disorders [53, 33, 17, 41].

The extensive bibliography accumulated has led to the approval of rTMS for the treatment of medication-resistant depression by both the Food and Drug Administration (United States of America) and the Federal Medical Council in Brazil.

1.2.4 The appearance of rTMS as a therapy for drug-resistant depression

Following the pioneering work of Pascual-Leone and colleagues [70], several other researchers have confirmed the efficacy of high-frequency stimulation over the left dorsolateral prefrontal cortex (DLPFC) [28, 83, 42], as well as low-frequency rTMS over the right DLPFC [8, 21, 12]. rTMS as a therapy for treatment-resistant depression was approved by the FDA (2011) as well as by the CFM (2011).

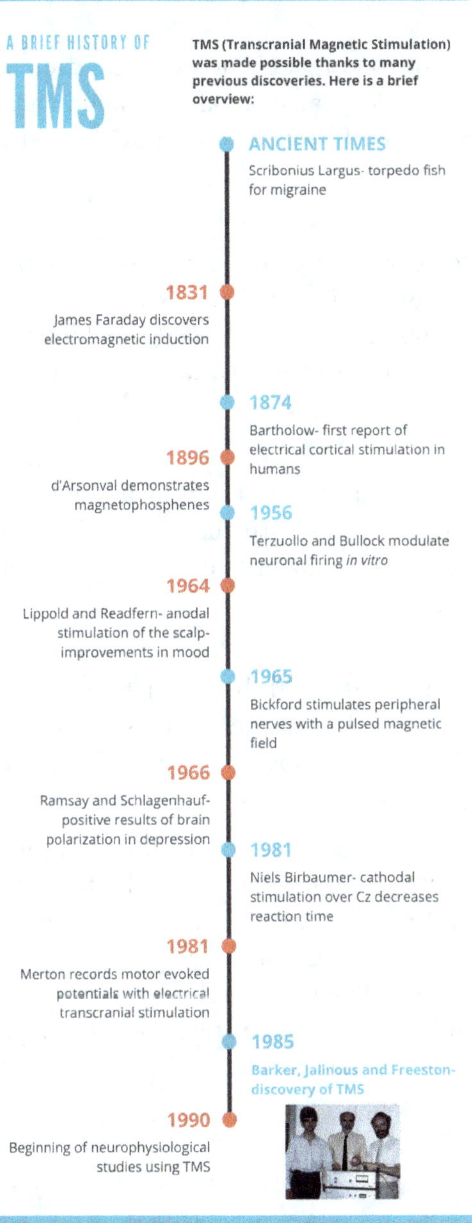

Figure 1.4: Timeline of TMS development

1.2.5 The rebirth of transcranial direct current stimulation (tDCS)

The success of studies on therapeutic neuromodulation with rTMS has led to a renewed interest in an older transcranial neuromodulation technique, namely, cerebral polarization, currently renamed as tDCS (transcranial direct current stimulation).

As early as 1964, Redfearn, Lippold, and Costain published a study on the effects of anodic stimulation of the scalp [81], reporting favorable results on patients' mood. In 1966, Ramsay and Schlagenhauf also reported favorable outcomes from the application of low-voltage direct current over the skulls of depressed patients [80].

In 1981, Elbert and Birbaumer reported a reduction in response time in humans in neuropsychological tasks with the application of cathodic current over the Cz position of the international 10-20 electrode placement system for electroencephalography [23].

In 1998, Priori and colleagues in Italy found that cerebral polarization with low-intensity currents applied over the intact skull was capable of modifying the excitability threshold of the motor cortex to rTMS [78].

In the year 2000, Michael Nitsche and colleagues in Germany conducted a series of experiments using tDCS in conjunction with TMS to verify if prior tDCS over the motor cortex would modulate the amplitudes of motor responses obtained subsequently with TMS from the same area [65]. These authors found that anodic stimulation increased the excitability of the motor cortex, reducing its response threshold to TMS; the opposite occurred when the cortex was previously exposed to cathodic tDCS. These results confirmed and expanded the findings of Priori et al [78].

Fregni et al. in 2005, also demonstrated positive effects of tDCS on working memory tests in normal volunteers, suggesting that other cortical areas, besides the motor ones, are amenable to neuromodulation by this technique [25].

Currently, several experimental studies are underway to verify the similarities and differences between the effects of neuromodulation produced by rTMS and tDCS, always considering that cathodic tDCS is generally inhibitory, resembling low-frequency rTMS in this aspect, and that anodic tDCS, excitatory, would be similar to high-frequency rTMS. It is worth noting, however, that tDCS has not yet been approved by the FDA or the CFM for clinical use. The greatest hopes for positive results concern depression and chronic pain.

2 Contributions of TMS to studies on brain plasticity

To assess the importance of visual experience we deprived kittens of vision in one eye at various times after birth, and for various lengths of time. In one kitten the lids of one eye were closed at an age of 9 weeks for a period of 1 month. While a number of nerve cells were driven from connection with both eyes, the balance was nevertheless still grossly abnormal, with an unusually large proportion of cells strongly preferring the normal eye... Finally, it was found that a few months of deprivation in an adult cat were not enough to produce any cortical deficit or any morphological changes in the geniculate. These experiments, then, indicate that very young kittens are particularly susceptible to the effects of deprivation, and that the susceptibility decreases with each month of life, possibly even vanishing in the mature animal.

— Hubel & Wiesel (1981 Nobel Prize for Medicine and Physiology)

The citation above describes the experiments on visual deprivation conducted on newborn and adult cats by Hubel and Wiesel in the 1970s and 1980s. Partly due to these experiments, for a long time biology held as dogma the absence of dynamic modifications (plasticity) in the adult brain; it was believed that the ability to drastically reorganize in response to injuries, for example, was an exclusive feature of the infant brain, and that there was a critical period for brain plasticity, after which this reorganization would no longer be possible.

However, subsequent studies would show that this rigid behavior of neuronal connections in adults is a peculiar characteristic of the visual system, related to the precision required of these connections for three-dimensional vision, and that in other circuits of the central nervous system there was no such critical period nor such almost absolute restriction of brain plasticity in adult animals.

In fact, several studies conducted in the 1980s and 1990s by Sanes, Merzenich, Pons, Kaas, among others [88, 76, 2, 43], revealed a degree of brain plasticity in adult animals that was previously unsuspected. These studies involved, basically, peripheral manipulations such as limb amputations, artificial syndactyly, nerve sections, etc., followed by the study of possible modifications in the somatotopic representation of body parts involved in the sensory cortex.

Studies in animals revealed that:

- When a limb is amputated, its somatotopic representation area in the sensory or motor cortex does not remain silent, but becomes related to the stump or other body regions represented in the adjacent cortex;
- When artificial syndactyly is performed, the digits (of the bat, for example) no longer have separate cortical representations, but share a single cortical area in the postcentral gyrus;
- These modifications can occur extremely rapidly.

However, studies on brain plasticity in animals, due to technical ease aspects, focused on sensory systems; the opposite occurred with studies in humans using TMS , since, parallel to animal studies, it became possible, with this technique, to non-invasively study the somatotopic representation areas in the precentral gyrus.

2.1 Motor cortex mapping with TMS

Motor cortex mapping can be performed by moving the coil over the scalp, simultaneously recording the muscle responses (MEPs - motor evoked potentials). The results can be presented in the form of a topographic map, with muscle responses of different amplitudes represented by peaks and valleys of different colors. With this technique, it was possible to:

- demonstrate the weak ipsilateral projections of the motor cortex to the upper limb muscles;
- reveal the bilateral and unilateral projections of the cranial nerve pairs;
- demonstrate that in chronic amputees there is an increase in the cortical

motor representation area of proximal body segments, i.e., of the stump;

■ likewise, to show that in paraplegics there is an increase in the cortical representation area of the abdominal muscles.

An exception to the trend of exclusive study of somesthetic areas in animals was the study by Sanes [88]. This researcher studied the effects of facial nerve section in rats. In the rat, the facial nerve innervates the vibrissae ("whiskers"), which are extremely important for this animal, providing tactile afferences and being capable of a wide range of movements.

In these experiments, the motor cortex was mapped before and after surgical section of the facial nerve. It was observed that, before nerve section, stimulation of the motor area for the vibrissa produced movements of that body part, while stimulation of an adjacent area produced movements of the front paw. However, a few minutes after nerve section, stimulation of the vibrissa area started to produce movements of the front paw, in the absence of any movement of the vibrissa, which was paralyzed; stimulation of the front paw representation area also continued to produce movements of this limb; as a result, the cortical motor representation area of the front paw increased greatly in size!

Sanes and colleagues postulated that the mechanism underlying this type of plasticity should be an "unmasking" of pre-existing synaptic connections between the cortical representation area of the vibrissa and the front paw. To test this hypothesis, these researchers perfused the motor cortex of animals with the intact facial nerve with a GABA inhibitor, bicuculline. In the motor cortex treated in this way, it was observed that stimulation of the somatotopic representation area of the vibrissa produced movements of both the vibrissa itself and the front paw, confirming the experimental hypothesis!

Would the rapid reorganization of cortical somatotopy that occurred in Sanes' rats also occur in humans? This possibility was tested in experiments of "simulated amputation" of the upper limb in humans, by Brasil-Neto and colleagues [11].

These authors used an anesthetic block of the forearm in normal volunteers (which acted as a "reversible amputation") and verified what would happen to the excitability of the cortical motor representation area of a muscle proximal to the block, the biceps. Within a few minutes of complete anesthesia installation, there was a large increase in the excitability of the cortical motor representation area of the biceps, which also quickly returned to normal with the cessation of anesthesia. It was thus proven that unmasking of pre-existing synaptic connections can also occur in humans, leading to an acute form of plasticity.

2.2 TMS in brain plasticity during motor learning

Pascual-Leone et al. [69] found an increase in the motor cortical representation area of the hand in novice individuals practicing piano scales in the laboratory; the cortical representation returned to its original size when the exercises were abandoned. A very interesting observation was that there was a certain increase in the hand area even with piano exercises imagined by the volunteers, without actual execution of the manual task.

In another study, the same group found that in blind Braille readers, the motor cortical representation area of the "reading" finger was always larger than that of the contralateral finger and also larger than that observed in the cortex of blind individuals who did not read Braille or of volunteers with normal vision [72]. Additionally, when these individuals took "reading vacations" for a few weeks, there was some reduction in the size of the somatotopic representation area of the "reading" finger! Another interesting observation on functional brain imaging of these subjects was that, when reading Braille, there was activation of the visual cortex [87], indicative of cross-modal plasticity. It was then shown that transient transcranial magnetic stimulation of the occipital (visual) cortex induced errors in Braille reading tasks and distorted the tactile perceptions of blind subjects [19].

Thus, we see that TMS, after its development, which aimed primarily at use in neurological diagnostics, allowing for the acquisition of motor evoked potentials useful in the assessment of demyelinating diseases of the central nervous system, such as multiple sclerosis, proved to be a very valuable tool for neurophysiology studies, especially in the field of brain plasticity.

3 Rationale for rTMS in the Treatment of Depression

3.1 Evidence of hypo- or hyperexcitability of specific cortical areas in depression

Neuroimaging studies show a reduction in metabolism in the left dorsolateral prefrontal cortex (DLPFC), contrasting with increased activity in the right DLPFC, in patients with depression [32, 45]. High-frequency rTMS to the left is an attempt to increase and normalize the excitability of the left DLPFC; the same reasoning applies to the use of inhibitory (low-frequency) rTMS directed to the right DLPFC. The lasting effects of rTMS treatment, even after the sessions have concluded, could come from the reestablishment of physiological mechanisms for maintaining normal levels of prefrontal cortical excitability. However, it is not possible to exclude mechanisms of cortical plasticity induction, such as phenomena similar to long-term potentiation, axonal sprouting, and even gene expression induction [93].

3.2 Evidence-based rTMS: large, multicenter, controlled studies

3.2.1 rTMS Studies in Depression

To date, at least three large, multicenter, randomized, double-blind, placebo-controlled studies have been published on the effects of rTMS in depression, including a total of 703 adult patients with major depressive disorder (MDD) who had not responded to 1 to 4 adequate attempts of medication treatment [73].

The first multicenter controlled study was published in 2007 [67] and included 20 centers in the United States, 2 in Australia, and one in Canada. In this study, patients randomized to receive active rTMS showed clinically significant improvement on the Montgomery-Asberg depression scale after 4 weeks, compared to those who received placebo stimulation.

The second multicenter, randomized, and placebo-controlled study was conducted by the NIH, National Institutes of Health (USA), independently of the industry [31]. The study was conducted at 4 American universities and included 190 outpatients with MDD. There was a significantly higher number of remissions in patients treated with active rTMS, according to the Hamilton depression scale (15% in the active rTMS group versus 4% in the placebo rTMS group, p<0.01).

The third study was the Brainsway study, which involved 13 American centers, 1 in Canada, 2 in Europe, and 4 in Israel [54]. Of 212 patients, 181 completed the study. This study included deep rTMS. The remission rates by the Hamilton scale were 32.6% versus 14.6% for active rTMS and placebo, respectively (p<0.01). For deep rTMS, these rates were 38.4% and 21.4%, respectively (p<0.014).

3.2.2 Approval of the Brainsway® Stimulator for Deep rTMS in Obsessive-Compulsive Disorder

In 2018, the FDA approved rTMS for the treatment of obsessive-compulsive disorder. The FDA's decision was based on a controlled clinical study of 100 patients with obsessive-compulsive disorder. Participants were randomized to receive 6 weeks of treatment with rTMS or placebo stimulation (sham). Five 25-minute sessions were conducted weekly. Before each session, a therapist employed a behavioral therapy known as tailored symptom provocation, which encouraged patients to think about their anxieties and/or compulsions.

One month after the last session, 38.1% of the patients in the rTMS group obtained a clinical response (defined as a reduction greater than 30% in symptom severity measured by the Yale-Brown scale), compared to 11.1% in the placebo group. Additionally, 54.8% of the patients in the rTMS group achieved a partial response (more than 20% reduction in symptom severity), compared to 26.7%

in the placebo group. However, it is important to note that this approval is valid only for the Brainsway® deep brain stimulation equipment.

4 How to Use rTMS in the Treatment of Depression

4.1 Identifying Patients with Pharmacoresistant Depression

The official indication for rTMS therapy in depression is as follows:

> Treatment of Major Depressive Disorder (MDD) in adult patients who have not achieved satisfactory improvement with previous antidepressant medication treatment during the current episode.
>
> Perera et al., 2016 [73]

Clinically, the patients for whom rTMS was indicated in the three main published randomized clinical trials [67, 31, 54] exhibited the following characteristics [73, 34]

■ Moderate to severe resistance in the current episode- patients had received 1-4 attempts of adequate antidepressant medication and 1 to 23 total attempts of antidepressant treatment. The majority of the Clinical TMS Society members consider an attempt of antidepressant medication treatment at an adequate dose for 6 to 8 weeks without results as a failure of pharmacological therapy. The period can be shorter in the case of intolerance to the medication.

■ A recurrent course of the disease is another criterion of resistance. More than 95% of the patients in the cited trials had experienced previous episodes. The average age of the patients was 49 years.

■ Moderate to severe illness with functional disability at the initial assessment, where reduced work productivity reflected the functional disability. More than 50% of the preatients were unemployed due to their illness, and 30% were receiving government disability benefits.

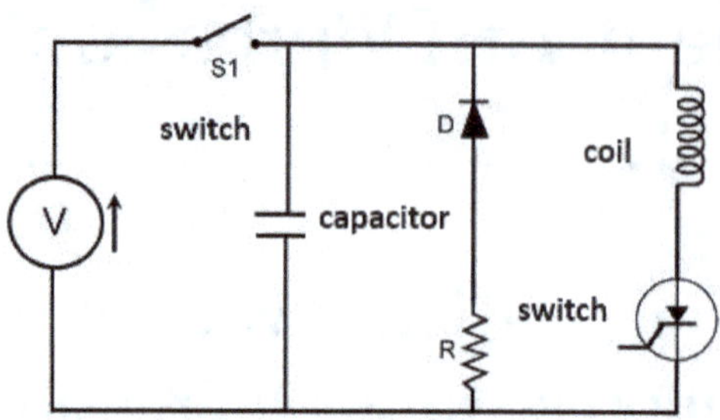

Figure 4.1: The basic circuit of a magnetic stimulator

4.2 Understanding the Magnetic Stimulator

The fundamental circuit of a magnetic stimulator is relatively simple, as shown in Figure 4.1: a powerful capacitor is coupled to a coil, and its discharge controlled by switches and resistors. When the switch is activated, the capacitor discharges, and electric current flows through the coil, creating, by Faraday's law of electromagnetic induction, a rapidly time-varying magnetic field, with an approximate intensity of 2 Tesla (similar in intensity to magnetic resonance imaging).

This magnetic field, in turn, is capable of inducing, through the intact skull, electrical currents deep within the cerebral parenchyma, stimulating axons situated parallel to the cortical surface (intracortical interneurons). This is where rTMS differs from transcranial electrical stimulation (TES), which depolarizes pyramidal neurons in the axon implantation zone.

4.2.1 Brief history of TMS development

Transcranial Magnetic Stimulation (TMS) was preceded by 5 years by the successful attempt to stimulate the motor cortex through the skull using electrical stimuli (TES) [58]. This type of high-voltage transcranial electrical stimulation had the drawback of causing significant pain by stimulating nociceptors in the scalp. However, the possibility of producing muscle contractions by stimulating the motor cortex (motor evoked potentials) suggested new possibilities for studying the physiology of the human motor cortex, as well as for developing new diagnostic tests for motor system pathologies.

The next step in the development of TMS was the demonstration, by Barker et al.[3], that it was possible to stimulate peripheral nerves and the motor cortex

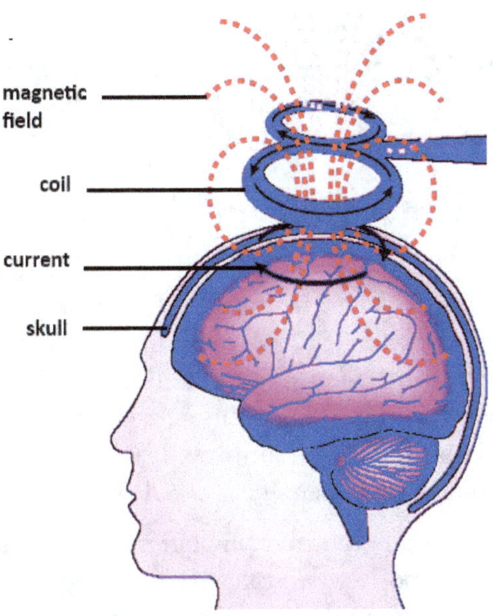

Figure 4.2: The magnetic field induces electrical currents in the cerebral tissue

with magnetic stimuli in a completely painless manner.

During TMS, a magnetic field is produced, with flux lines passing perpendicularly to the plane of the coil, which is generally positioned tangentially to the scalp, as in Figure 4.2. The magnetic field can reach an intensity of 2 Tesla, with a duration of approximately 100 µseconds. Neural elements that are parallel to the magnetic field are stimulated more efficiently at points where the intensity of the field changes as a function of distance. If the neural elements are not completely parallel to the magnetic field, the most effective stimulation occurs at the bending points of these elements.

The significant advantage of rTMS over TES is that there is no pain or discomfort since there is no accumulation of electrical charges on the scalp. Skin and bone are transparent to the magnetic field, and stimulation can even occur without the coil touching the skull, although the magnitude of the magnetic field rapidly decreases with distance.

Transcranial electrical stimulation initially stimulates the axon's implantation zone, resulting in the so-called D waves (direct wave). With both TES and TMS, subsequent I waves (indirect waves) occur, which result from the trans-synaptic activation of intracortical interneurons. With TMS, at most stimulation intensities and coil angles, D responses are not produced. Due to this difference in the physiology of motor neuron activation, the muscle responses obtained with TES have a slightly shorter latency than those obtained by TMS [22].

Magnetic stimulators can operate with two different types of classic coils: the figure-8 (or "butterfly") coil and the circular coil. The main advantage of the figure-8 coil is that it produces more focal stimulation, concentrated at the intersection of the "wings" of the "butterfly". In contrast, the circular coil, when positioned over the vertex, for example, stimulates both cerebral hemispheres. However, it is possible to preferentially stimulate the left cerebral hemisphere by using the coil in a position where the current flows counter-clockwise; for preferential right-side stimulation, simply turn the coil to the opposite side. The direction of current flow is usually marked on modern devices.

Magnetic pulses can have different morphologies, namely: monophasic or biphasic. Biphasic pulses have positive and negative components and constitute a sinusoidal wave; they are commonly used. Some devices are capable of producing monophasic pulses, generally used in experimental protocols.

One problem with magnetic stimulation is the rapid decrease in the intensity of the magnetic field with distance, making cortical stimulation relatively superficial. A special type of coil has been developed to allow for deeper penetration of the magnetic field: the "H" coil [103].

A recent innovation is deep brain stimulation devices using coils of this type, which are not yet approved by the CFM (Federal Council of Medicine in Brazil) for clinical use and will not be discussed here.

4.3 Commercially Available Magnetic Stimulators

Some of the most well-known commercial magnetic stimulation devices include Magandmore®, Magstim®, MagVenture®, Neurosoft®, and Neurostar®. Equipment for rTMS with a special coil that allows for deeper stimulation is also commercially available.

It is important to verify, at the time of purchase, whether the device in question is registered with ANVISA (Brazil's National Health Surveillance Agency), to legally use it in Brazil. More information about these devices can be obtained by visiting their respective websites:

- Brainsway®: http://www.brainsway.com

- Magandmore®: http://www.magandmore.com

- Magstim®: http://www.magstim.com

- Magventure®:http://www.magventure.com

- Neurosoft®: http://www.neurosoft.com.br

- Neurostar®: https://neurostar.com

4.4 Minimum Requirements for the rTMS Room

The resolution 1.986/2012 of the Brazilian Federal Council of Medicine established minimum requirements for the operation of an rTMS room, aiming at the prompt management of any possible complications. In practice, the most predictable complication, although rare, would be the occurrence of a seizure. This complication is likely to be more probable with the use of deep transcranial magnetic stimulation with the Brainsway® device. In clinical trials, the risks for each individual patient were: 0.1% for NeuroStar and 0.12% for Brainsway® [73].

Nonetheless, the following items are currently required by the CFM:

1. Oxygen point;
2. Pulse oximeter;
3. Venturi mask;
4. Laryngeal mask;
5. Nasal cannula, mask for nebulization;
6. Laryngoscope (handle and at least one curved and one straight blade);
7. Stylet;
8. Tubes for orotracheal intubation of different sizes;
9. Bag valve mask (Ambu);
10. Scalpels, syringes, and needles for drug administration;
11. Adhesive tape;
12. Suction device (portable);
13. Personal protective equipment (gloves, glasses, etc.)

Besides the above material, the CFM resolution also lists some mandatory medications:

1. Analgesics;
2. Injectable and oral diazepam;
3. Injectable phenobarbital;
4. Injectable hydantoin;
5. Injectable midazolam;
6. Antiarrhythmics;
7. Bronchodilators;
8. 0.9% saline solution;
9. 25% and 50% glucose solution.

Currently, there is at least one Brazilian commercial business offering this complete kit, on its website (https://www.casadatecnologiamedica.com.br/). It is also important to use ear plugs for both the patient and the applicator to prevent possible, though unlikely, acoustic trauma from the coil discharge noise [59].

Finally, it is advisable to avoid unnecessary exposure to the electromagnetic field by the rTMS applicator. A recent Swedish study demonstrated, with a Magventure® device, an electromagnetic field level above what is established as acceptable by European standards, up to 40 cm away from the coil [63].

Although to date no proven deleterious effects of exposure to electromagnetic fields are known, it is advisable to avoid excessive chronic exposure by the applicator trying to maintain a distance greater than 40 cm from all body parts that can be moved away from the coil (the hand and arm of the applicator will obviously always be exposed if the mechanical coil holder provided by most manufacturers is not used). The safety of the patient depends on adherence to protocols with the standards already established as acceptable and with lower probabilities of inducing adverse events, such as the maximum recommended duration, in seconds, of the rTMS trains [101, 85, 84], according to Table 4.1.

A major drawback of rTMS treatment of depression is the delay in treatment response. Recently, various protocols for accelerated rTMS have been proposed, delivering more than one daily session. Theta-burst stimulation has been used for this purpose [39]. However, more research work still needs to be done before accelerated rTMS can be routinely used [100].

Frequency (Hz)	100%	110%	120%
1	1,800	1,800	360
5	10	10	10
10	5	5	4.2
20	2.05	1.6	1
25	1.28	0.84	0.4

Table 4.1: Maximal duration of TMS trains, in seconds, taking into account the frequency and intensity of stimulation

5 Patient Interview

5.1 Patient Interview

5.1.1 Selection for rTMS

As previously explained, patients must meet the criteria for pharmacotherapy-resistant MDD (Major Depressive Disorder). They should be subjected to a preliminary screening questionnaire to identify possible contraindications to the procedure:

Preliminary Questionnaire for Transcranial Magnetic Stimulation

1. Do you have epilepsy or have ever had a seizure?
2. Have you ever had fainting or syncope? If yes, please describe on what occasion or occasions?
3. Have you ever had a head injury that was diagnosed as a concussion or was associated with a loss of consciousness?
4. Do you have hearing problems or ringing in the ears?
5. Do you have a cochlear implant?
6. Are you pregnant or is there any chance that you might be?
7. Do you have metal in your brain, skull, or any other part of your body (for example, fragments, clips, plates, etc.)? If yes, describe the type of metal.
8. Do you have an implanted neurostimulator (for example, DBS, epidural, subdural, vagal stimulator)?
9. Do you have a pacemaker or intra-cardiac catheters?
10. Are you carrying a medication infusion pump?
11. Are you taking any medications? Please make a list.
12. Have you undergone rTMS in the past? If yes, were there any issues?
13. Have you undergone MRI in the past? If yes, were there any issues?

Adapted from Rossi et al. 2009 [85]

Assessment Instruments for MDD

Regarding the instruments for assessing depression, most members of the Clinical TMS Society, in a recent survey, pointed out the following [73]:

- Patient Health Questionnaire-9 (PHQ-9);
- Beck Depression Inventory.

Patient agreement: Informed Consent Form (ICF)

Before the procedure, the patient must be informed about the possible risks and benefits of the treatment. Each clinic should develop its own ICF, in the national language and in accessible form, avoiding technical terms that might confuse the patient. It is interesting that this Term is signed by at least one witness.

Additionally, to instruct the patient about all the details of the procedure, oral presentations and videos can be used. In some situations, it is also interesting to discuss the procedure with the patient's family members.

6 The Procedure

6.1 Procedures that should occur sequentially

1. Localization of the motor point (hot spot) for the right abductor pollicis brevis (APB) muscle;
2. Determination of the motor threshold (MT);
3. Determination of F3;
4. Fixation of the coil over the left DLPFC;
5. Administration of the treatment protocol.

6.1.1 Localization of the Motor Point (Hot Spot) for the Right APB

■ The patient should be seated comfortably and relaxed, with the right arm supported in supination to observe any contraction of the thenar eminence and/or thumb; it is important to keep the muscle relaxed to avoid facilitation of responses to cortical stimulation [10, 34].

■ The coil is positioned 4 cm laterally and 1 cm anterior to the vertex, with the handle shifted 45° in relation to the sagittal plane [10], over the approximate position C3 of the international 10-20 system for EEG electrode placement;

■ Using this point as the origin, the coil is moved over the scalp until an unequivocal movement of the APB is observed;

■ The coil is moved 0.5 cm in the four cardinal directions, triggering 3 stimuli at each point; if none of these points produce a better response, the origin is considered the motor point (hot spot for the APB). If one of the points produces a better response, it will be considered the new origin and the process is repeated.

6.1.2 Determination of the Motor Threshold (MT)

■ Once the hot spot for the APB is determined, 10 magnetic stimuli are fired at the intensity used for the hot spot determination, always ensuring that at least 5 contractions occur in the right thenar eminence;

■ The intensity is reduced by 2%, another 10 stimuli are fired, and the process is repeated until fewer than 5 contractions are observed in a series of 10 stimuli;

■ When no longer observing at least 5 contractions in 10 stimuli, the motor threshold (MT) will be 2% above this intensity;

■ The intensity to be used in the treatment will be determined based on this MT value, depending on the protocol used. For example, if the patient's MT is 40% of the device's maximum power and the protocol prescribes 120% of the motor threshold, then the intensity used will be 48%. [1] [2]

6.1.3 Determination of F3

Having determined the intensity to be used, it is necessary to reposition and fix the coil over the F3 position of the 10-20 system, which corresponds to the left DLPFC - some authors simply advocate moving the coil 5 cm anteriorly along a parasagittal line after determining the motor point (5 cm rule), but this procedure often results in poor positioning of the coil in relation to the DLPFC, especially in patients with larger skulls [61].

The F3 position can be determined in the usual way, with skull measurements routinely used in electroencephalography, but this is a laborious procedure. An alternative is to use the Beam F3 system [4]. This system uses a mathematical formula that allows, from just 3 skull measurements, to locate F3 taking into account the cephalic dimensions of each patient. Versions of the Beam F3 system software are available for free download (Mac and Windows) at http://www.clinicalresearcher.org. There is also an online platform to perform this calculation without the need to install any computer program: http://clinicalresearcher.org/F3.

A study compared the left DLPFC location obtained by the Beam F3 system with that provided by MRI neuronavigation and revealed good concordance [61]. The online calculation platform of the Beam F3 system also provides a value adjusted to this small difference observed in neuron-

---

[1] Instead of using the visual criterion, the motor evoked potential can be recorded on devices equipped with an electromyograph; in this case, instead of counting movements of the thenar eminence, electromyographic responses with amplitude greater than 50 μvolts peak-to-peak are counted

[2] A study highlighted the possibility that the MT is overestimated when using the method of visual detection of muscle contractions [102]. On average, the MT detected by the visual method was 11.3% higher than that determined electromyographically. This should be taken into account to avoid excessive stimulation intensities during treatments, which would increase the risks of adverse events.

Parameter	O'Reardon	CFM 1	CFM 2
Intensity (% MT)	120	110	120
Frequency	10 Hz	10 Hz	5 Hz
Series duration	4 seg	5 seg	10 seg
Interval between series	26 seg	25 seg	20 seg
Number of series	75	25	20
Pulses per session	3,000	1,250	1,250

Table 6.1: Classical O'Reardon protocol and protocols suggested by CFM 1.986/2012 resolution

avigation, which allows for even greater precision in the positioning of the coil.

6.1.4 Fixation of the Coil over the Left DLPFC

Finally, having determined the intensity and location of the magnetic stimuli, we must fix the coil over the skull, which can be done with the use of the adjustable mechanical fixator that accompanies many devices. In the absence of this resource, the operator must hold the coil firmly during each series of stimuli, taking great care to avoid modifications in the positioning and angle of the coil in relation to the scalp.

6.1.5 Administration of the Treatment Protocol

High-frequency rTMS over the left DLPFC

Level of evidence "A" (proven efficacy) from the European panel [51, 50]. After fixing the coil, the execution of the generally pre-programmed treatment protocol on the computer connected to the device is triggered. There are differences regarding the classic protocol that led to the approval of TMS by the US FDA [67] and that was also followed by the NIH multicentric study [31] and those suggested by the CFM, according to the Table 6.1.

Low-frequency rTMS over the right DLPFC

Level of evidence "B" (probable efficacy) from the European panel [51, 50]. The CFM also suggests a low-frequency rTMS protocol over the right DLPFC, with the following characteristics: 1 Hz frequency, 80 to 100% of the motor threshold intensity, in a single series of 20 minutes per session. Here it is important to note: the dosing of rTMS has been continuously

reevaluated in the literature, as studies develop; as mentioned above, there
are devices that already come with pre-programmed software, but these
must also be critically reassessed in light of the most updated literature.

6.1.6 Possible Side Effects

As previously mentioned, the main potential complication of rTMS con-
cerns the induction of a seizure, which can occur in 1 out of 30,000 treatment sessions ($< 0.003\%$) [73]. Vasovagal syncope may occur during rTMS
and should be differentiated from a convulsive episode [82]. The following characteristics help differentiate between vasovagal syncope and ictal
episode:

■ History of previous episodes: during blood collection, vaccination, after
prolonged orthostatism, etc.;

■ Recovery of consciousness in seconds;

■ Absence of: tonic or clonic movements, tongue biting, urinary inconti-
nence, prolonged loss of consciousness, and postictal confusional state;

■ Episodes not preceded by sleep deprivation, exposure to strobe lights,
or ingestion of alcoholic beverages.[3]

6.2 Evaluating the Results of rTMS

6.2.1 Documentation of Effect

Objective documentation of clinical benefit should be obtained routinely
in every clinical rTMS service. This is important for clinical assessment
and may also support possible reimbursement by health insurance [73].
Self-assessment scales by the patient, such as the PHQ-9 [89], the IDS-R,
or the Beck Depression Inventory, can be used.

6.2.2 Post-treatment Planning

Once the maximum benefit is achieved, rTMS treatment should be gradually discontinued, and a maintenance regimen should be prescribed for
each patient. In the pioneering studies by Neuronetics® and the NIH,
rTMS was gradually withdrawn over a period of 3 weeks (3 sessions per
week, then 2, and finally 1 in the last week). After suspending rTMS,
patients were kept on therapy with just one medication, and new access to
rTMS was allowed in case of symptom relapse [73].

---

[3]A common side effect is headache, which is usually mild and responds well to common
analgesics

6.2.3 Use of rTMS Combined with Medications

rTMS can be administered in the presence or absence of antidepressant medications, and currently, there is no evidence of an increased risk of complications due to this combination. However, as medications can interfere with the motor threshold (MT), in case of medication change, this parameter should be reevaluated before new rTMS sessions [73].

6.2.4 Patients Who Do Not Respond to rTMS

For patients who do not respond to 4 to 6 weeks of rTMS treatment, most members of the Clinical rTMS Society prescribe one or two additional weeks of daily rTMS, definitively suspending treatment if there is no response. A smaller number of members definitively end treatment already at 6 weeks without result. In patients with partial response after the sixth week, most members extend the same protocol or modify the dose, target, or frequency of applications [73].

7 Applications

7.1 Applications Approved by the Federal Council of Medicine

7.1.1 Auditory Hallucinations in Schizophrenia

The CFM Resolution 1.986/2012 validated the use of low-frequency rTMS for the treatment of auditory hallucinations (AH) in schizophrenia. The first description of the potential utility of rTMS in AH was published in 1999 [38]. Since then, various studies have been published, with success in some patients and failure in others, which seems to be due to the peculiarities of each case [62].

One of the factors considered likely for the variability of the results is the location of the left temporo-parietal junction, which is the desired target of rTMS. The clinical location of this point on the scalp is usually midway between T3 and P3 (international 10-20 system of electrode positioning for electroencephalography). However, at least one study comparing this clinical positioning with that obtained by sophisticated functional magnetic resonance neuronavigation techniques showed no significant differences between the two approaches [92].

Recent literature reviews have concluded that rTMS, especially when applied over the left temporo-parietal region with a frequency of 1 Hz, is effective for the treatment of AH, including in patients with medication-resistant AH [91][37]. A recent meta-analysis suggested that women and young people show the best responses [46].

Suggested Parameters:

- Evidence level "C" (possible efficacy) by the European panel [51, 50].

- Coil positioning: midway between T3 and P3

- Frequency: less than or equal to 1 Hz (CFM: 1 Hz)

- Intensity: 80 to 100% of MT (CFM)

- Duration: 20 minutes (CFM)

- Number of series: 1 (CFM)

7.1.2 Preoperative Motor Mapping

Transcranial magnetic stimulation can be used in neurosurgical planning as a non-invasive tool for motor cortex mapping. This mapping provides safety to the neurosurgeon during surgeries for epilepsy or tumors, for example, indicating which cortical areas can be excised without significant risk of postoperative motor deficits [24].

When coupled with stereotactic neuronavigation techniques and the patient's own transcranial magnetic resonance imaging, TMS can be as efficient as invasive cortical mapping techniques [75].

7.1.3 Preoperative Language Mapping

The possibility of using rTMS as a tool for localizing Broca's area was first envisioned with the initial description of speech arrest produced in normal individuals by rTMS [68]. A recent study described good concordance between pre-surgical language mapping with rTMS and intraoperative mapping in awake patients [74].

7.2 Experimental rTMS in Other Disorders

Besides its well-established role in the treatment of MDD, rTMS has been the subject of several studies showing encouraging results in various neurological diseases. However, it is worth noting that, in Brazil, the CFM has not yet been consulted and has not published any resolution regarding these additional applications of rTMS.

7.2.1 Epilepsy

Low-frequency rTMS (less than or equal to 1 Hz) can induce a reduction in cortical excitability, probably via gabaergic modulation, and has achieved, in some studies, a reduction in the number of seizures in patients with refractory epilepsy [86].

It is important to remember that rTMS's ability to influence the focus is limited by its difficulty in reaching deeper cortical areas, such as in temporal lobe epilepsy. The best results were obtained with epilepsy with more superficial neocortical foci [27]. However, the only controlled and randomized clinical trial of low-frequency rTMS for epilepsy published to date [90] involved 7 patients with neocortical foci and was unable to demonstrate benefits of rTMS in this clinical condition.

Suggested Parameters:

■ Evidence level: "C" (possible efficacy) confirmed by the European panel for low-frequency rTMS [50].

■ Coil type: butterfly

■ Coil positioning: over the probable neocortical focus

■ Frequency: less than or equal to 1 Hz

■ Intensity: 90% of MT [15]

- Number of sessions: 5 to 14
- Pulses per session: 900 to 1,500

7.2.2 Post-Stroke Rehabilitation

After a stroke, there appears to be an imbalance between the excitability of the motor cortex of the lesioned hemisphere and that of the contralateral hemisphere. In other words, the transcallosal inhibition normally exerted by the healthy hemisphere on the affected one becomes relatively excessive.

Thus, two different rTMS neuromodulation strategies have been advocated: excitatory (high-frequency) stimulation over the ipsilateral motor cortex to the stroke or inhibitory (low-frequency) stimulation over the contralateral motor cortex [64].

Another potential use in stroke would be in accelerating the recovery of aphasia when associated with speech therapy [60]. In these cases too, the described strategy is to inhibit the right homologous area to Broca's area with low-frequency rTMS, or to stimulate Broca's area itself, with high frequency.

Suggested Parameters:

Inhibitory protocol:

- Evidence level "A" (proven efficacy) by the European panel [50] for stroke with 1 week to 6 months of evolution, at least for the recovery of hand motor function
- Coil type: butterfly
- Coil positioning: over the motor area (C3 or C4) of the contralesional hemisphere (in the case of aphasia, right homolog of Broca's area)
- Frequency: less than or equal to 1 Hz

- Intensity: 80-120% of MT

- Number of sessions: 5 to 15

- Pulses per session: 200 to 1,000

Excitatory protocol:

- Evidence level "B" (probable efficacy) by the European panel [50] for stroke with 1 week to 6 months of evolution, at least for the upper limb motor function

- Coil type: butterfly

- Coil positioning: over the motor area (C3 or C4) of the ipsilesional hemisphere (in the case of aphasia, over Broca's area)

- Frequency: 5 to 10 Hz

- Intensity: 90-130% of MT

- Number of sessions: 5 to 15

- Pulses per session: 200 to 1,000

Inhibitory protocol for stroke rehabilitation in the chronic phase (after 6 months from the event)

- Evidence level "C" (possible efficacy) by the European panel [50] for stroke with 6 months or more of evolution, at least for the recovery of hand motor function [36].

7.2.3 Parkinson's Disease

rTMS applied over the motor cortex and the DLPFC induces dopamine release in the striatum [94]. A recent meta-analysis of the literature concluded that high-frequency rTMS over the motor cortex appears to be more efficient in improving motor symptoms and signs of Parkinson's disease [66]. This same meta-analysis concluded that low-frequency rTMS does not appear to have a significant effect on these signs and symptoms.

Studies have also been published with low-frequency rTMS applied over the supplementary motor area or motor cortex with a reduction in levodopa-induced dyskinesias [24, 66].

Suggested Parameters:

Motor signs and symptoms of Parkinson's disease

- Evidence level "B" (probable efficacy) by the European panel [51, 50]
- Coil type: butterfly
- Coil positioning: over both motor cortices
- Intensity: 80 to 120% of MT
- Frequency: 5 to 25 Hz
- Number of sessions: 6 to 10
- Pulses per session: 600 to 3,000

Note: the effects are modest and likely without clinical significance

Levodopa-induced dyskinesias

- No evidence level classification by the European panel [51, 50]
- Coil type: butterfly
- Coil positioning: over the motor cortex or supplementary motor area
- Frequency: less than or equal to 1 Hz

Depression in Parkinson's Disease

■ Evidence level "B" (probable efficacy) by the European panel [51, 50]

■ Same protocol as for depression in patients without Parkinson's disease

7.2.4 Dystonias

Since dystonias involve a failure in intracortical inhibition mechanisms, it seems reasonable to expect an improvement in symptomatology with low-frequency rTMS over the motor cortex and premotor areas. Thus, the main studies published to date have targeted these areas, in addition to the supplementary motor area [34].

Suggested Parameters:

■ No evidence level classification by the European panel [51, 50]

■ Coil type: butterfly

■ Coil positioning: over the motor cortex, premotor areas, or supplementary motor area

■ Frequency: less than or equal to 1 Hz

7.2.5 Chronic Neuropathic Pain

The first report of chronic pain control by motor cortex stimulation was published over three decades ago [99]. Since then, reports of chronic pain control by motor cortex stimulation have been consistently published, many using invasive methods [9]. Neurosurgical implantation of epidural or subdural electrodes seems to give the best results, but recently, several successful reports of rTMS have been published [44]. Positive results have been reported for neuropathic pain, fibromyalgia, complex regional pain syndrome, and visceral pain [24].

Suggested Parameters:

- Evidence level "A" (proven efficacy) by the European panel [50]

- Coil positioning: over the contralateral motor cortex to the pain

- Frequency: 5 Hz to 20 Hz

- Intensity: below MT (80-90% of MT)

- Number of sessions: induction: 10 consecutive daily sessions, excluding weekends; maintenance: 3 sessions with a one-week interval, 3 sessions with a 15-day interval, and 3 sessions with a one-month interval [44]

There is evidence that stimulation of the hand representation area contralateral to the pain is effective regardless of the location of neuropathic pain and that there is a cumulative effect of multiple sessions [50].

7.2.6 Tinnitus

Idiopathic tinnitus is common in the population (approximately 10%), but its treatment is difficult. Several small controlled studies have demonstrated that low-frequency rTMS applied over the primary auditory cortex seems to reduce symptoms, but the ideal parameters have not yet been established [48, 24].

Given the only moderate results and the discrepancy between different studies, there is currently a trend towards the adoption of individualized treatment protocols, after tests of acute improvement of tinnitus with prefrontal or temporo-parietal stimulation with different frequencies, to identify the best treatment strategy for each patient [47].

Suggested Parameters:

- Evidence level "C" (possible efficacy) by the European panel [51, 50]

- Coil type: butterfly

- Coil positioning: over the left primary auditory cortex or contralateral to the "buzzing"

- Frequency: 1 Hz

- Intensity: 80 to 120% of MT

- Number of sessions: 5 to 20

- Pulses per session: 900 to 2,000

7.2.7 Panic Disorder

The results of different studies are conflicting, and the studies are small. The following parameters are based on a recent article with promising results [56]. A recent review on rTMS in Anxiety Disorders was published [41].

A recent study [1] showed beneficial results from high-frequency stimulation over the right dorsolateral prefrontal cortex, and the latest publication by the European panel assigned a recommendation level B to this treatment modality.

Suggested Parameters:

- Evidence level "B" (probable efficacy) by the European panel [50]

- Coil type: butterfly

- Coil positioning: right DLPFC

- Frequency: 20 Hz

- Intensity: 100% of MT

- Number of sessions: 2 to 3 days a week, for 4 weeks

- Pulses per session: 2,400

8 Transcranial direct current stimulation

8.1 Suggested Applications for Transcranial Direct Current Stimulation (tDCS)

8.1.1 Transcranial direct current stimulation- a brief history

Studies on transcranial direct current stimulation (tDCS) began in the 1960s with animal experimentation [6, 79]. In 1964, Bindman et al. published a study demonstrating that resting membrane potentials could be altered by applying low-intensity currents to the cortical areas of rats, and that this effect could last for hours after the stimulation had ended [6].

Purpura and McMurtry [79] observed an increase in spontaneous neuronal excitability following the application of anodic tDCS to cells in the pyramidal tract of cats, and the opposite effect with cathodic stimulation. Since that decade, the goal of applications was to improve learning and memory and to treat neuropsychiatric diseases. However, therapeutic interest in tDCS waned due to technological limitations of the time and advances made by the pharmaceutical industry [77].

Interest in the application of low-intensity galvanic current transcranially resurged in the 1990s, with structured protocols. Priori et al. [78] demonstrated, through transcranial magnetic stimulation (TMS), that low-intensity cathodic tDCS decreased motor cortex excitability in humans. Nitsche and Paulus [65] conducted a similar study, proving, through excitability tests using TMS, the polarity-dependent modulatory effect of tDCS on the motor cortex of healthy individuals.

8.1.2 Mechanism of Action, Technique, and Equipment

Unlike TMS, tDCS does not cause depolarization of the cortical neuronal membrane or trigger action potentials. tDCS induces modifications in the resting membrane potential, thus bringing this potential closer to or further from the action potential threshold [65]. The effectiveness of tDCS in inducing acute changes in neuronal membrane polarity depends on the current density (which determines the magnitude of the induced electric field) and is a function of current intensity and electrode size . Generally, these parameters are: intensity of 1 or 2 mA and electrodes of 35 cm^2 in published studies.

The stimulation typically employs conductive rubber electrodes of 35 cm^2 wrapped in synthetic sponge soaked in saline solution and fixed with an elastic Velcro® band. The active electrode is classically placed over the area of interest according to the international 10-20 system (the same used for electroencephalography), with the reference electrode over the contralateral supraorbital region.

As of this writing, the following transcranial direct current stimulation devices are registered with the Brazilian agency ANVISA:

- DC–stimulator by Neuroconn® GMBH (Germany)
- Microstim electrostimulator by NKL® (Brazil)
- Transcranial Direct Current Stimulator by Soterix® (USA)
- Transcranial Direct Current Neurostim® by Medssuply Electro-electronics Ltda (Brazil)

8.1.3 Recent Efficacy Evidence

Parallel to research with TMS, various applications of tDCS have been proposed for numerous neuropsychiatric conditions. Two recent literature reviews deserve special mention: that of the European panel [52] and that by Fregni et al. [26]. In the latter, two treatment protocols with tDCS had their efficacy reclassified to level "A" (proven efficacy) or "B" (probable efficacy). These are: pain conditions and major depressive disorder.

8.1.4 Pain

Suggested parameters

■ Level of evidence "C" (possible efficacy) from the European panel [52]

■ Level of evidence "B" (probable efficacy) from the review by Fregni et al. [26] for neuropathic pain, fibromyalgia, migraine, and postoperative pain

■ Electrode placement: M1 (C3/C4 electrode of the 10-20 system) contralateral for focal or lateralized pain; M1 of the dominant cerebral hemisphere for diffuse pain; left dorsolateral prefrontal cortex; primary visual cortex (V1) for migraine

■ Stimulus polarity: anodic

■ Electrode size: 35 cm^2

■ Intensity: 2 mA

■ Session duration: 10 to 20 minutes

■ Number of sessions: 20

8.1.5 Major Depressive Disorder

Suggested parameters

■ Level of evidence "B" (probable efficacy) from the European panel [52] and "A" (proven efficacy) from the review by Fregni et al. [26]

■ Electrode placement: anode over the left DLPFC (F3) and cathode over the contralateral supraorbital region

■ Electrode size: 35 cm^2

- Intensity: 2 mA
- Session duration: 20 to 30 minutes
- Number of sessions: more than 10

9 Challenges and Future Perspectives

9.1 Challenges and Future Perspectives for the Use of Therapeutic rTMS in Neuropsychiatric Diseases

9.1.1 Need for More Controlled Studies

With the aim of modulating cortical excitability in an inhibitory or excitatory manner, transcranial magnetic stimulation (TMS) has been tested in various clinical contexts where hypo- or hyperexcitability of the cortex exists.

The confirmation of the effects and clinical efficacy of TMS treatments, in each case, depends on the same clinical, controlled, and multicentric studies that are conducted for pharmacological therapies.

To put into perspective the general problems related to the design, research, and future perspectives of these therapies, let us examine the case of epilepsy, which is the prototype neurological condition where cortical excitability is altered, making it a perfect candidate for non-pharmacological neuromodulatory therapies.

9.1.2 The Case of Epilepsy

Why attempt rTMS in epilepsy? According to the review published by Chen et al. [16]:

■ Out of 1,795 patients treated for epilepsy for the first time, 1,144 remained seizure-free for a year or more;

■ This represents only 63.7% of all patients;

■ More than one-third of patients remain with uncontrolled seizures with pharmacological treatment;

■ The plethora of new drugs introduced in the last 20 years has not substantially changed this scenario.

However, the first attempt to use rTMS in epilepsy was not for therapeutic purposes but diagnostic: in 1990, Andreas Hufnagel and colleagues [40] proposed using TMS as a tool to activate epileptic foci in the pre-surgical evaluation of patients undergoing surgery for refractory epilepsy. However, these authors used single pulses of TMS and were not able to activate irritative foci, which could reduce the time needed for their pre-operative localization.

The first study published as an attempt to reduce the number of seizures by treating the focus with inhibitory, low-frequency rTMS was that of Frithjof Tergau and colleagues, in 1999 [95]. These authors based their research on studies that demonstrated the inhibitory effects of low-frequency rTMS in animals. It was an open pilot study, involving 9 patients, and showed a reduction in the number of seizures in the weeks following treatment.

William Theodore, at the National Institutes of Health (NIH), in 2002, conducted the first controlled study [96], with 24 patients with refractory epilepsy, and described a small, short-lived effect of seizure reduction with low-frequency rTMS. The authors emphasized that the best results were obtained with neocortical foci, probably because the intensity of the magnetic field drops rapidly with distance from the coil, and deeper foci may not be reached by rTMS stimuli.

In 2006, the Brazilian Felipe Fregni and colleagues published a controlled, double-blind study [27], in which they treated 21 patients with cortical development malformations and refractory epilepsy and found significant reductions in seizure frequency, by 58% in the treated group.

In 2007, Roberto Cantello and colleagues also conducted a double-blind, placebo-controlled study [13] of 43 patients with refractory epilepsy and found no significant effect on seizure frequency; however, they reported a significant effect in reducing epileptiform abnormalities on the electroencephalogram in interictal periods.

In 2016, the only controlled and randomized clinical trial of low-frequency rTMS for epilepsy published to date [90] involved only 7 patients with neocortical foci and was not able to demonstrate benefits of rTMS in this clinical condition.

9.1.3 rTMS and Status Epilepticus

Attempts to use rTMS to abort status epilepticus episodes were recently reviewed by Zeiler et al.[104]. This author concluded that:

a) There is little data in the literature on generalized status epilepticus;
b) In focal status epilepticus, the response rate is 80%, and in refractory focal status epilepticus, 50%;
c) Therefore, there is a need for more controlled, multicentric studies with a large number of patients to approve or reject rTMS as a valid therapeutic modality in epilepsy.

9.1.4 A Possible Future Perspective: Use in ICU

Both animal studies and clinical trials conducted so far demonstrate a frequent normalization of the interictal electroencephalogram with the use of low-frequency rTMS. The clinical relevance of this finding is still uncertain, but rTMS could be used experimentally, for example, in cases of non-convulsive status epilepticus (NCSE), which often go unnoticed in ICUs and are a significant cause of consciousness alterations in severely ill patients with various clinical conditions. These patients are not necessarily sufferers of epilepsy or neurological diseases, and even antibiotics, such as cefepime, can induce NCSE [49].

With the progressive introduction of continuous brain monitoring by digital electroencephalogram, currently advocated by LaRoche and Haider [49], among others, with consequent better detection of NCSE cases, an opportunity may arise for studies on non-invasive neuromodulation by rTMS in these patients. As they are generally exposed to multiple drugs and their side effects, treating NCSE with rTMS, if proven efficient, without the use of additional drugs, would potentially be safer in these cases.

9.1.5 Need for Research Sponsors

In the case of pharmaceuticals, pharmaceutical companies usually finance large controlled multicentric studies. For rTMS, the proof of its efficacy in depression depended on studies that received financial support from equipment manufacturers, as well as studies funded by state agencies, such as the National Institutes of Health (NIH), in the USA.

In the future, this trend is expected to continue, and it will be necessary for foundations, governments, and the industry to join efforts to conduct the necessary research to validate the routine use of therapeutic neuromodulation by rTMS in various neuropsychiatric diseases.

10 Conclusion

Repetitive Transcranial Magnetic Stimulation (rTMS) constitutes a non-invasive therapeutic modality, devoid of systemic effects, already approved by the U.S. Food and Drug Administration (FDA) and the Brazilian Federal Council of Medicine (CFM) for the treatment of major depressive disorder.

While the CFM has sanctioned its use for auditory hallucinations in schizophrenia, this application has only received a "C" level of evidence (possible efficacy) from the European panel of experts [51, 50].

Other applications of rTMS are in the final stages of efficacy proof and may likewise be approved for clinical use: chronic pain and stroke rehabilitation. As a relatively new technique, further studies are required to assess its clinical utility across a broad spectrum of neuropsychiatric disorders.

Finally, the technology continues to evolve, and new modalities of rTMS, featuring different coils and physical characteristics of stimulation, are under constant development. Moreover, another non-invasive neuromodulation modality, transcranial Direct Current Stimulation (tDCS), is still in the phase of intense experimentation and has not yet received CFM endorsement for clinical use. However, the recent reclassification of its effect to an "A" level of evidence for depression and "B" for various types of pain suggests that this neuromodulation modality may soon also be approved for clinical use.

The information contained herein should be evaluated in the context of an extremely dynamic field of study, which is rapidly advancing with new publications and technological refinements.

The Author Joaquim P. Brasil-Neto

Dr. Joaquim Pereira Brasil-Neto is a neurologist and clinical neurophysiologist, a Full Member of both the Brazilian Academy of Neurology and the Brazilian Society of Clinical Neurophysiology. He was among the pioneering researchers in the field of transcranial magnetic stimulation in the early 1990s. He has published works alongside notable scholars such as Mark Hallett, Leonardo Cohen, Álvaro Pascual-Leone, Eric Wasserman, and Josep Valls-Solé, among others. He began his research career as a Visiting Fellow at the National Institutes of Neurological Disorders and Stroke (USA).

Currently, he serves as a faculty member and researcher at the Centro Universitário Unieuro, where his primary research focuses include transcranial magnetic stimulation and transcranial direct current stimulation.

Bibliography

[1] M.-J. Ahmadizadeh and M. Rezaei. Unilateral right and bilateral dorsolateral prefrontal cortex transcranial magnetic stimulation in treatment post-traumatic stress disorder: A randomized controlled study. *Brain Research Bulletin*, 140:334–340, June 2018.

[2] T. Allard, S. A. Clark, W. M. Jenkins, and M. M. Merzenich. Reorganization of somatosensory area 3b representations in adult owl monkeys after digital syndactyly. *Journal of Neurophysiology*, 66(3):1048–1058, Sept. 1991.

[3] A. T. Barker, R. Jalinous, and I. L. Freeston. Non-invasive magnetic stimulation of human motor cortex. *The Lancet*, 325(8437):1106–1107, 1985. Publisher: Elsevier.

[4] W. Beam, J. J. Borckardt, S. T. Reeves, and M. S. George. An efficient and accurate new method for locating the F3 position for prefrontal TMS applications. *Brain Stimul*, 2:50–54, Jan. 2009.

[5] R. G. Bickford. Neuronal stimulation by pulsed magnetic fields in animals and man. In *Dig. 6th Int. Conf. Med. Electronics Biol. Eng (Tokyo), 1965*, 1965.

[6] L. J. Bindman, O. C. J. Lippold, and J. W. T. Redfearn. The action of brief polarizing currents on the cerebral cortex of the rat (1) during current flow and (2) in the production of long-lasting after-effects. *The Journal of Physiology*, 172(3):369–382, Aug. 1964.

[7] T. V. Bliss and T. Lømo. Long-lasting potentiation of synaptic transmission in the dentate area of the anaesthetized rabbit following stimulation of the perforant path. *The Journal of physiology*, 232(2):331–356, 1973. Publisher: Wiley Online Library.

[8] R. Boechat-Barros and J. P. Brasil-Neto. Transcranial Magnetic Stimulation in depression: results of bi-weekly treatment. *Brazilian Journal of Psychiatry*, 26:100–102, 2004. Publisher: SciELO Brasil.

[9] J. P. Brasil-Neto. Motor Cortex Stimulation for Pain Relief: Do Corollary Discharges Play a Role? *Frontiers in human neuroscience*, 10:323, 2016.

[10] J. P. Brasil-Neto, L. G. Cohen, M. Panizza, J. Nilsson, B. J. Roth, and M. Hallett. Optimal focal transcranial magnetic activation of the human motor cortex: effects of coil orientation, shape of the induced current pulse, and stimulus intensity. *Journal of Clinical Neurophysiology: Official Publication of the American Electroencephalographic Society*, 9(1):132–136, Jan. 1992.

[11] J. P. Brasil-Neto, J. Valls-Solé, A. Pascual-Leone, A. Cammarota, V. E. Amassian, R. Cracco, P. Maccabee, J. Cracco, M. Hallett, and L. G. Cohen. Rapid modulation of human cortical motor outputs following ischaemic nerve block. *Brain: A Journal of Neurology*, 116 (Pt 3):511–525, June 1993.

[12] A. R. Brunoni, A. Chaimani, A. H. Moffa, L. B. Razza, W. F. Gattaz, Z. J. Daskalakis, and A. F. Carvalho. Repetitive transcranial magnetic stimulation for the acute treatment of major depressive episodes: a systematic review with network meta-analysis. *JAMA psychiatry*, 74(2):143–152, 2017. Publisher: American Medical Association.

[13] R. Cantello and others. Slow repetitive TMS for drug-resistant epilepsy: clinical and EEG findings of a placebo-controlled trial. *Epilepsia*, 48(2):366–374, 2007.

[14] U. Cerletti. Electroshock therapy. *Journal of Clinical & Experimental Psychopathology*, 1954.

[15] B. S. Chang. TMS: A Tailored Method of Stimulation for Refractory Focal Epilepsy? *Epilepsy Curr*, 13:162–163, July 2013.

[16] Z. Chen, M. J. Brodie, D. Liew, and P. Kwan. Treatment Outcomes in Patients With Newly Diagnosed Epilepsy Treated With Established and New Antiepileptic Drugs: A 30-Year Longitudinal Cohort Study. *JAMA neurology*, 75(3):279–286, Mar. 2018.

[17] C. Chung and M. Mak. Effect of repetitive transcranial magnetic stimulation on physical function and motor signs in Parkinson's disease: a systematic review and meta-analysis. *Brain stimulation*, 9(4):475–487, 2016. Publisher: Elsevier.

[18] L. G. Cohen, S. Bandinelli, H. R. Topka, P. Fuhr, B. J. Roth, and M. Hallett. Topographic maps of human motor cortex in normal and pathological conditions: mirror movements, amputations and spinal cord injuries. *Electroencephalography and Clinical Neurophysiology-Supplements only*, 43:36–50, 1994. Publisher: Amsterdam: Elsevier, c1999-.

[19] L. G. Cohen, P. Celnik, A. Pascual-Leone, B. Corwell, L. Faiz, J. Dambrosia, M. Honda, N. Sadato, C. Gerloff, M. D. Catala´, and M. Hallett. Functional relevance of cross-modal plasticity in blind humans. *Nature*, 389(6647):180–183, Sept. 1997. Publisher: Nature Publishing Group.

[20] L. G. Cohen, B. J. Roth, E. M. Wassermann, H. Topka, P. Fuhr, J. Schultz, and M. Hallett. Magnetic stimulation of the human cerebral cortex, an indicator of reorganization in motor pathways in certain pathological conditions. *Journal of clinical neurophysiology: official publication of the American Electroencephalographic Society*, 8(1):56–65, 1991.

[21] B. Dell'Osso, L. Oldani, G. Camuri, C. Dobrea, L. Cremaschi, B. Benatti, C. Arici, B. Grancini, and A. C. Altamura. Augmentative repetitive Transcranial Magnetic Stimulation (rTMS) in the acute treatment of poor responder depressed patients: a comparison study between high and low frequency stimulation. *European Psychiatry*, 30(2):271–276, 2015. Publisher: Cambridge University Press.

[22] V. Di Lazzaro, A. Oliviero, P. Profice, E. Saturno, F. Pilato, A. Insola, P. Mazzone, P. Tonali, and J. C. Rothwell. Comparison of descending volleys evoked by

transcranial magnetic and electric stimulation in conscious humans. *Electroencephalography and Clinical Neurophysiology*, 109(5):397–401, Oct. 1998.

[23] T. Elbert, W. Lutzenberger, B. Rockstroh, and N. Birbaumer. The influence of low-level transcortical DC-currents on response speed in humans. *International Journal of Neuroscience*, 14(1-2):101–114, 1981. Publisher: Taylor & Francis.

[24] M. C. Eldaief, D. Z. Press, and A. Pascual-Leone. Transcranial magnetic stimulation in neurology: A review of established and prospective applications. *Neurol Clin Pract*, 3:519–526, Dec. 2013.

[25] F. Fregni, P. S. Boggio, M. Nitsche, F. Bermpohl, A. Antal, E. Feredoes, M. A. Marcolin, S. P. Rigonatti, M. T. A. Silva, W. Paulus, and A. Pascual-Leone. Anodal transcranial direct current stimulation of prefrontal cortex enhances working memory. *Exp Brain Res*, 166(1):23–30, 2005.

[26] F. Fregni, M. M. El-Hagrassy, K. Pacheco-Barrios, S. Carvalho, J. Leite, M. Simis, J. Brunelin, E. M. Nakamura-Palacios, P. Marangolo, G. Venkatasubramanian, D. San-Juan, W. Caumo, M. Bikson, A. R. Brunoni, and Neuromodulation Center Working Group. Evidence-Based Guidelines and Secondary Meta-Analysis for the Use of Transcranial Direct Current Stimulation in Neurological and Psychiatric Disorders. *The International Journal of Neuropsychopharmacology*, 24(4):256–313, Apr. 2021.

[27] F. Fregni, P. T. M. Otachi, A. Valle, P. S. Boggio, G. Thut, S. P. Rogonatti, A. Pascual-Leone, and K. D. Valente. A randomized clinical trial of repetitive transcranial magnetic stimulation in patients with refractory epilepsy. *Ann Neurol*, 60:447–455, 2006.

[28] F. Fregni, C. Santos, M. Myczkowski, R. Rigolino, J. Gallucci-Neto, E. Barbosa, K. Valente, A. Pascual-Leone, and M. Marcolin. Repetitive transcranial magnetic stimulation is as effective as fluoxetine in the treatment of depression in patients with Parkinson's disease. *Journal of Neurology, Neurosurgery & Psychiatry*, 75(8):1171–1174, 2004. Publisher: BMJ Publishing Group Ltd.

[29] L. Geddes. The history of magnetophosphenes. *IEEE Engineering in Medicine and Biology Magazine*, (4):101–102, 2008.

[30] M. S. George, T. A. Ketter, and R. M. Post. Prefrontal cortex dysfunction in clinical depression. *Depression*, 2(2):59–72, 1994. Publisher: Wiley Online Library.

[31] M. S. George, S. H. Lisanby, D. Avery, W. M. McDonald, V. Durkalski, M. Pavlicova, B. Anderson, Z. Nahas, P. Bulow, P. Zarkowski, and others. Daily left prefrontal transcranial magnetic stimulation therapy for major depressive disorder: a sham-controlled randomized trial. *Archives of general psychiatry*, 67(5):507–516, 2010. Publisher: American Medical Association.

[32] M. S. George, E. M. Wassermann, W. A. Williams, A. Callahan, T. A. Ketter, P. Basser, M. Hallett, and R. M. Post. Daily repetitive transcranial magnetic

stimulation (rTMS) improves mood in depression. *NeuroReport*, 6(14):1853, Oct. 1995.

[33] P. V. O. Gomes, J. P. Brasil-Neto, N. Allam, and E. Rodrigues de Souza. A randomized, double-blind trial of repetitive transcranial magnetic stimulation in obsessive-compulsive disorder with three-month follow-up. *The Journal of Neuropsychiatry and Clinical Neurosciences*, 24(4):437–443, 2012.

[34] M. Hallett. Transcranial magnetic stimulation: a primer. *Neuron*, 55:187–199, July 2007.

[35] L. J. Harris and J. B. Almerigi. Probing the human brain with stimulating electrodes: the story of Roberts Bartholow's (1874) experiment on Mary Rafferty. *Brain and Cognition*, 70(1):92–115, 2009. Publisher: Elsevier.

[36] R. L. Harvey and others. Randomized Sham-Controlled Trial of Navigated Repetitive Transcranial Magnetic Stimulation for Motor Recovery in Stroke: The NICHE Trial. *Stroke*, 49(9):2138–2146, 2018.

[37] H. He, J. Lu, L. Yang, J. Zheng, F. Gao, Y. Zhai, J. Feng, Y. Fan, and X. Ma. Repetitive transcranial magnetic stimulation for treating the symptoms of schizophrenia: A PRISMA compliant meta-analysis. *Clinical Neurophysiology*, 128(5):716–724, May 2017.

[38] R. E. Hoffman, N. N. Boutros, R. M. Berman, E. Roessler, A. Belger, J. H. Krystal, and D. S. Charney. Transcranial magnetic stimulation of left temporoparietal cortex in three patients reporting hallucinated 'voices.'. *Biological psychiatry*, 46:130–132, 1999.

[39] Y.-Z. Huang, M. J. Edwards, E. Rounis, K. P. Bhatia, and J. C. Rothwell. Theta Burst Stimulation of the Human Motor Cortex. *Neuron*, 45(2):201–206, Jan. 2005.

[40] A. Hufnagel, C. E. Elger, H. F. Durwen, D. K. Böker, and W. Entzian. Activation of the epileptic focus by transcranial magnetic stimulation of the human brain. *Ann Neurol*, 27:49–60, 1990.

[41] A. Iannone, A. P. d. M. Cruz, J. P. Brasil-Neto, and R. Boechat-Barros. Transcranial magnetic stimulation and transcranial direct current stimulation appear to be safe neuromodulatory techniques useful in the treatment of anxiety disorders and other neuropsychiatric disorders. *Arquivos de neuro-psiquiatria*, 74:829–835, 2016. Publisher: SciELO Brasil.

[42] P. G. Janicak and M. E. Dokucu. Transcranial magnetic stimulation for the treatment of major depression. *Neuropsychiatric disease and treatment*, 11:1549, 2015. Publisher: Dove Press.

[43] J. H. Kaas. The reorganization of somatosensory and motor cortex after peripheral nerve or spinal cord injury in primates. *Progress in Brain Research*, 128:173–179, 2000.

[44] M. M. Klein, R. Treister, T. Raij, A. Pascual-Leone, L. Park, T. Nurmikko, F. Lenz, and others. Transcranial magnetic stimulation of the brain: guidelines for pain treatment research. *Pain*, 156:1601, 2015.

[45] M. Koenigs and J. Grafman. The functional neuroanatomy of depression: distinct roles for ventromedial and dorsolateral prefrontal cortex. *Behavioural brain research*, 201(2):239–243, 2009. Publisher: Elsevier.

[46] S. Koops, C. W. Slotema, C. Kos, L. Bais, A. Aleman, J. D. Blom, and I. E. C. Sommer. Predicting response to rTMS for auditory hallucinations: Younger patients and females do better. *Schizophrenia Research*, 195:583–584, May 2018.

[47] P. M. Kreuzer, T. B. Poeppl, R. Rupprecht, V. Vielsmeier, A. Lehner, B. Langguth, and M. Schecklmann. Individualized Repetitive Transcranial Magnetic Stimulation Treatment in Chronic Tinnitus? *Frontiers in Neurology*, 8:126, Apr. 2017.

[48] B. Langguth, D. de Ridder, J. L. Dornhoffer, P. Eichhammer, R. L. Folmer, E. Frank, F. Fregni, C. Gerloff, E. Khedr, T. Kleinjung, M. Landgrebe, S. Lee, J.-P. Lefaucheur, A. Londero, R. Marcondes, A. R. Moller, A. Pascual-Leone, C. Plewnia, S. Rossi, T. Sanchez, P. Sand, W. Schlee, D. Pysch, T. Steffens, P. van de Heyning, and G. Hajak. Controversy: Does repetitive transcranial magnetic stimulation/ transcranial direct current stimulation show efficacy in treating tinnitus patients? *Brain Stimul*, 1(3):192–205, July 2008.

[49] S. M. LaRoche and M. D. H. A. Haider, editors. *Handbook of ICU EEG monitoring*. Springer, Publishing Company, 2018.

[50] J.-P. Lefaucheur, A. Aleman, C. Baeken, D. H. Benninger, J. Brunelin, V. Di Lazzaro, S. R. Filipović, C. Grefkes, A. Hasan, F. C. Hummel, S. K. Jääskeläinen, B. Langguth, L. Leocani, A. Londero, R. Nardone, J.-P. Nguyen, T. Nyffeler, A. J. Oliveira-Maia, A. Oliviero, F. Padberg, U. Palm, W. Paulus, E. Poulet, A. Quartarone, F. Rachid, I. Rektorová, S. Rossi, H. Sahlsten, M. Schecklmann, D. Szekely, and U. Ziemann. Evidence-based guidelines on the therapeutic use of repetitive transcranial magnetic stimulation (rTMS): An update (2014-2018). *Clinical Neurophysiology: Official Journal of the International Federation of Clinical Neurophysiology*, 131(2):474–528, Feb. 2020.

[51] J.-P. Lefaucheur, N. André-Obadia, A. Antal, S. S. Ayache, C. Baeken, D. H. Benninger, R. M. Cantello, M. Cincotta, M. de Carvalho, D. De Ridder, H. Devanne, V. Di Lazzaro, S. R. Filipović, F. C. Hummel, S. K. Jääskeläinen, V. K. Kimiskidis, G. Koch, B. Langguth, T. Nyffeler, A. Oliviero, F. Padberg, E. Poulet, S. Rossi, P. M. Rossini, J. C. Rothwell, C. Schönfeldt-Lecuona, H. R. Siebner, C. W. Slotema, C. J. Stagg, J. Valls-Sole, U. Ziemann, W. Paulus, and L. Garcia-Larrea. Evidence-based guidelines on the therapeutic use of repetitive transcranial magnetic stimulation (rTMS). *Clinical Neurophysiology*, 125(11):2150–2206, Nov. 2014.

[52] J.-P. Lefaucheur, A. Antal, S. S. Ayache, D. H. Benninger, J. Brunelin, F. Co-giamanian, M. Cotelli, D. De Ridder, R. Ferrucci, B. Langguth, P. Marangolo, V. Mylius, M. A. Nitsche, F. Padberg, U. Palm, E. Poulet, A. Priori, S. Rossi, M. Schecklmann, S. Vanneste, U. Ziemann, L. Garcia-Larrea, and W. Paulus. Evidence-based guidelines on the therapeutic use of transcranial direct current stimulation (tDCS). *Clinical Neurophysiology: Official Journal of the International Federation of Clinical Neurophysiology*, 128(1):56–92, Jan. 2017.

[53] J.-P. Lefaucheur, X. Drouot, Y. Keravel, and J.-P. Nguyen. Pain relief induced by repetitive transcranial magnetic stimulation of precentral cortex. *Neuroreport*, 12(13):2963–2965, 2001. Publisher: LWW.

[54] Y. Levkovitz, M. Isserles, F. Padberg, S. H. Lisanby, A. Bystritsky, G. Xia, A. Tendler, Z. J. Daskalakis, J. L. Winston, P. Dannon, and others. Efficacy and safety of deep transcranial magnetic stimulation for major depression: a prospective multicenter randomized controlled trial. *World Psychiatry*, 14(1):64–73, 2015. Publisher: Wiley Online Library.

[55] R. C. Malenka. The role of postsynaptic calcium in the induction of long-term potentiation. *Molecular neurobiology*, 5(2):289–295, 1991. Publisher: Springer.

[56] A. Mantovani, M. Aly, Y. Dagan, A. Allart, and S. H. Lisanby. Randomized sham controlled trial of repetitive transcranial magnetic stimulation to the dorsolateral prefrontal cortex for the treatment of panic disorder with comorbid major depression. *Journal of affective disorders*, 144:153–159, 2013.

[57] N. Mayr, C. Baumgartner, J. Zeitlhofer, and L. Deecke. The sensitivity of transcranial cortical magnetic stimulation in detecting pyramidal tract lesions in clinically definite multiple sclerosis. *Neurology*, 41(4):566–566, 1991. Publisher: AAN Enterprises.

[58] P. A. Merton and H. B. Morton. Stimulation of the cerebral cortex in the intact human subject. *Nature*, 285(5762):227, May 1980.

[59] C.-C. Miguel Angel, M.-M. Ignacio, C. G. Cordero G, T.-M. René, S.-Z. Mario, R.-G. Matilde, and G.-A. Adalberto. Transcranial magnetic stimulation and acoustic trauma or hearing loss in children. *Neurological research*, 23(4):343–346, 2001. Publisher: Taylor & Francis.

[60] C. Miniussi, S. F. Cappa, L. G. Cohen, A. Floel, F. Fregni, M. A. Nitsche, M. Oliveri, A. Pascual-Leone, W. Paulus, A. Priori, and V. Walsh. Efficacy of repetitive transcranial magnetic stimulation/transcranial direct current stimulation in cognitive neurorehabilitation. *Brain Stimul*, 1(4):326–336, 2008.

[61] A. Mir-Moghtadaei, R. Caballero, P. Fried, M. D. Fox, K. Lee, P. Giacobbe, Z. J. Daskalakis, D. M. Blumberger, and J. Downar. Concordance Between BeamF3 and MRI-neuronavigated Target Sites for Repetitive Transcranial Magnetic Stimulation of the Left Dorsolateral Prefrontal Cortex. *Brain Stimul*, 8:965–973, 2015.

[62] P. Moseley, C. Fernyhough, and A. Ellison. Auditory verbal hallucinations as atypical inner speech monitoring, and the potential of neurostimulation as a treatment option. *Neuroscience & Biobehavioral Reviews*, 37(10):2794–2805, 2013. Publisher: Elsevier.

[63] O. J. Møllerløkken, H. Stavang, and K. Hansson Mild. Staff exposure to pulsed magnetic fields during depression treatment with transcranial magnetic stimulation. *International journal of occupational safety and ergonomics*, 23(1):139–142, 2017. Publisher: Taylor & Francis.

[64] U. Najib, S. Bashir, D. Edwards, A. Rotenberg, and A. Pascual-Leone. Transcranial brain stimulation: clinical applications and future directions. *Neurosurg. Clin. N. Am*, 22:233–251, Apr. 2011.

[65] M. A. Nitsche and W. Paulus. Excitability changes induced in the human motor cortex by weak transcranial direct current stimulation. *The Journal of physiology*, 527(3):633–639, 2000. Publisher: Wiley Online Library.

[66] I. Obeso, A. Cerasa, and A. Quattrone. The Effectiveness of Transcranial Brain Stimulation in Improving Clinical Signs of Hyperkinetic Movement Disorders. *Frontiers in Neuroscience*, 9:486, 2015.

[67] J. P. O'Reardon, H. B. Solvason, P. G. Janicak, S. Sampson, K. E. Isenberg, Z. Nahas, W. M. McDonald, D. Avery, P. B. Fitzgerald, C. Loo, and others. Efficacy and safety of transcranial magnetic stimulation in the acute treatment of major depression: a multisite randomized controlled trial. *Biological psychiatry*, 62(11):1208–1216, 2007. Publisher: Elsevier.

[68] A. Pascual-Leone, J. R. Gates, and A. Dhuna. Induction of speech arrest and counting errors with rapid-rate transcranial magnetic stimulation. *Neurology*, 41:697–702, 1991.

[69] A. Pascual-Leone, D. Nguyet, L. G. Cohen, J. P. Brasil-Neto, A. Cammarota, and M. Hallett. Modulation of muscle responses evoked by transcranial magnetic stimulation during the acquisition of new fine motor skills. *Journal of Neurophysiology*, 74(3):1037–1045, Sept. 1995.

[70] A. Pascual-Leone, B. Rubio, F. Pallardó, and M. D. Catalá. Rapid-rate transcranial magnetic stimulation of left dorsolateral prefrontal cortex in drug-resistant depression. *The Lancet*, 348(9022):233–237, 1996. Publisher: Elsevier.

[71] A. Pascual-Leone, J. M. Tormos, J. Keenan, F. Tarazona, C. Cañete, and M. D. Catalá. Study and modulation of human cortical excitability with transcranial magnetic stimulation. *Journal of Clinical Neurophysiology*, 15(4):333–343, 1998. Publisher: LWW.

[72] A. Pascual-Leone and F. Torres. Plasticity of the sensorimotor cortex representation of the reading finger in Braille readers. *Brain*, 116(1):39–52, 1993. Publisher: Oxford University Press.

[73] T. Perera, M. S. George, G. Grammer, P. G. Janicak, A. Pascual-Leone, and T. S. Wirecki. The clinical TMS society consensus review and treatment recommendations for TMS therapy for major depressive disorder. *Brain stimulation*, 9(3):336–346, 2016. Publisher: Elsevier.

[74] T. Picht, S. M. Krieg, N. Sollmann, J. R"osler, B. Niraula, T. Neuvonen, P. Savolainen, and others. A comparison of language mapping by preoperative navigated transcranial magnetic stimulation and direct cortical stimulation during awake surgery. *Neurosurgery*, 72:808–819, 2013.

[75] T. Picht, S. Schmidt, S. Brandt, D. Frey, H. Hannula, T. Neuvonen, J. Karhu, P. Vajkoczy, and O. Suess. Preoperative functional mapping for rolandic brain tumor surgery: comparison of navigated transcranial magnetic stimulation to direct cortical stimulation. *Neurosurgery*, 69:581–589, 2011.

[76] T. P. Pons, P. E. Garraghty, A. K. Ommaya, J. H. Kaas, E. Taub, and M. Mishkin. Massive cortical reorganization after sensory deafferentation in adult macaques. *Science (New York, N.Y.)*, 252(5014):1857–1860, June 1991.

[77] A. Priori. Brain polarization in humans: a reappraisal of an old tool for prolonged non-invasive modulation of brain excitability. *Clinical Neurophysiology*, 114(4):589–595, Apr. 2003.

[78] A. Priori, A. Berardelli, S. Rona, N. Accornero, and M. Manfredi. Polarization of the human motor cortex through the scalp. *Neuroreport*, 9(10):2257–2260, 1998. Publisher: LWW.

[79] D. P. Purpura and J. G. McMurtry. Intracellular activities and evoked potential changes during polarization of motor cortex. *Journal of Neurophysiology*, 28(1):166–185, Jan. 1965. Publisher: American Physiological Society.

[80] J. Ramsay and G. Schlagenhauf. Treatment of depression with low voltage direct current. *Southern medical journal*, 59(8):932–934, 1966.

[81] J. W. T. Redfearn, O. C. J. Lippold, and R. Costain. A preliminary account of the clinical effects of polarizing the brain in certain psychiatric disorders. *British Journal of Psychiatry*, 110:773–785, 1964.

[82] P. Riedel, F. M. Korb, T. Karcz, and M. N. Smolka. Witnessing loss of consciousness during TMS–Syncope in contrast to seizure. *Clinical Neurophysiology Practice*, 1:58–61, 2016.

[83] M. A. Rosa, W. F. Gattaz, A. Pascual-Leone, F. Fregni, M. O. Rosa, D. O. Rumi, M. Myczkowski, M. F. Silva, C. Mansur, S. P. Rigonatti, and others. Comparison of repetitive transcranial magnetic stimulation and electroconvulsive therapy in unipolar non-psychotic refractory depression: a randomized, single-blind study. *International Journal of Neuropsychopharmacology*, 9(6):667–676, 2006. Publisher: Cambridge University Press Cambridge, UK.

[84] S. Rossi, A. Antal, S. Bestmann, M. Bikson, C. Brewer, J. Brockmöller, L. L. Carpenter, M. Cincotta, R. Chen, J. D. Daskalakis, V. Di Lazzaro, M. D. Fox, M. S. George, D. Gilbert, V. K. Kimiskidis, G. Koch, R. J. Ilmoniemi, J. P. Lefaucheur, L. Leocani, S. H. Lisanby, C. Miniussi, F. Padberg, A. Pascual-Leone, W. Paulus, A. V. Peterchev, A. Quartarone, A. Rotenberg, J. Rothwell, P. M. Rossini, E. Santarnecchi, M. M. Shafi, H. R. Siebner, Y. Ugawa, E. M. Wassermann, A. Zangen, U. Ziemann, and M. Hallett. Safety and recommendations for TMS use in healthy subjects and patient populations, with updates on training, ethical and regulatory issues: Expert Guidelines. *Clinical Neurophysiology*, 132(1):269–306, Jan. 2021.

[85] S. Rossi, M. Hallett, P. M. Rossini, A. Pascual-Leone, S. o. T. C. Group, and others. Safety, ethical considerations, and application guidelines for the use of transcranial magnetic stimulation in clinical practice and research. *Clinical neurophysiology*, 120(12):2008–2039, 2009. Publisher: Elsevier.

[86] A. Rotenberg. Prospects for clinical applications of transcranial magnetic stimulation and real-time EEG in epilepsy. *Brain Topogr*, 22:257–266, Jan. 2010.

[87] N. Sadato, A. Pascual-Leone, J. Grafman, V. Ibañez, M.-P. Deiber, G. Dold, and M. Hallett. Activation of the primary visual cortex by Braille reading in blind subjects. *Nature*, 380(6574):526–528, Apr. 1996. Publisher: Nature Publishing Group.

[88] J. N. Sanes, S. Suner, J. F. Lando, and J. P. Donoghue. Rapid reorganization of adult rat motor cortex somatic representation patterns after motor nerve injury. *Proceedings of the National Academy of Sciences of the United States of America*, 85(6):2003–2007, Mar. 1988.

[89] I. S. Santos, B. F. Tavares, T. N. Munhoz, L. S. P. d. Almeida, N. T. B. d. Silva, B. D. Tams, A. M. Patella, and A. Matijasevich. Sensibilidade e especificidade do Patient Health Questionnaire-9 (PHQ-9) entre adultos da população geral. *Cadernos de Saúde Pública*, 29:1533–1543, 2013. Publisher: SciELO Brasil.

[90] L. Seynaeve, A. Devroye, P. Dupont, and W. Van Paesschen. Randomized crossover sham-controlled clinical trial of targeted low-frequency transcranial magnetic stimulation comparing a figure-8 and a round coil to treat refractory neocortical epilepsy. *Epilepsia*, 57(1):141–150, Jan. 2016.

[91] C. W. Slotema, J. D. Blom, R. v. Lutterveld, H. W. Hoek, and I. E. Sommer. Review of the efficacy of transcranial magnetic stimulation for auditory verbal hallucinations. *Biol. Psychiatry*, 76:101–110, July 2014.

[92] C. W. Slotema, J. D. Blom, A. D. d. Weijer, K. M. Diederen, R. Goekoop, J. Looijestijn, K. Daalman, and others. Can low-frequency repetitive transcranial magnetic stimulation really relieve medication-resistant auditory verbal hallucinations? Negative results from a large randomized controlled trial. *Biol. Psychiatry*, 69:450–456, Mar. 2011.

[93] M. Stock, B. Kirchner, D. Waibler, D. Cowley, M. Pfaffl, and R. Kuehn. Effect of magnetic stimulation on the gene expression profile of in vitro cultured neural cells. *Neuroscience letters*, 526(2):122–127, 2012. Publisher: Elsevier.

[94] A. P. Strafella, T. Paus, M. Fraraccio, and A. Dagher. Striatal dopamine release induced by repetitive transcranial magnetic stimulation of the human motor cortex. *Brain*, 126:2609–2615, Dec. 2003.

[95] F. Tergau and others. Low-frequency repetitive transcranial magnetic stimulation improves intractable epilepsy. *The Lancet*, 353.9171:2209, 1999.

[96] W. H. Theodore and others. Transcranial magnetic stimulation for the treatment of seizures: a controlled study. *Neurology*, 59(4):560–562, 2002.

[97] H. Topka, L. G. Cohen, R. A. Cole, and M. Hallett. Reorganization of corticospinal pathways following spinal cord injury. *Neurology*, 41(8):1276–1276, 1991. Publisher: AAN Enterprises.

[98] G. Tsoucalas, M. Karamanou, M. Lymperi, V. Gennimata, and G. Androutsos. The "torpedo" effect in medicine. *International maritime health*, 65(2):65–67, 2014.

[99] T. Tsubokawa, Y. Katayama, T. Yamamoto, T. Hirayama, and S. Koyama. Chronic motor cortex stimulation for the treatment of central pain. In Springer, editor, *Advances in Stereotactic and Functional Neurosurgery 9*, pages 137–139. 1991.

[100] S. J. H. van Rooij, A. R. Arulpragasam, W. M. McDonald, and N. S. Philip. Accelerated TMS - moving quickly into the future of depression treatment. *Neuropsychopharmacology*, 49(1):128–137, Jan. 2024.

[101] E. M. Wassermann. Risk and safety of repetitive transcranial magnetic stimulation: report and suggested guidelines from the International Workshop on the Safety of Repetitive Transcranial Magnetic Stimulation, June 5–7, 1996. *Electroencephalography and Clinical Neurophysiology/Evoked Potentials Section*, 108(1):1–16, 1998. Publisher: Elsevier.

[102] G. G. Westin, B. D. Bassi, S. H. Lisanby, and B. Luber. Determination of motor threshold using visual observation overestimates transcranial magnetic stimulation dosage: safety implications. *Clin Neurophysiol*, 125:142–147, Jan. 2014.

[103] A. Zangen, Y. Roth, B. Voller, and M. Hallett. Transcranial magnetic stimulation of deep brain regions: evidence for efficacy of the H-coil. *Clinical Neurophysiology: Official Journal of the International Federation of Clinical Neurophysiology*, 116(4):775–779, Apr. 2005.

[104] F. A. Zeiler and others. Transcranial magnetic stimulation for status epilepticus. *Epilepsy research and treatment*, 2015.

Index

www.ingramcontent.com/pod-product-compliance
Lightning Source LLC
Chambersburg PA
CBHW050044260726
48658CB00005B/1757